Mastering Essential Math Skills

Whole Numbers and Integers

Richard W. Fisher

Whole Numbers and Integers - Item #404
ISBN 10: 0-9666211-4-X • ISBN 13: 978-0-9666211-4-3

Notes to the Teacher or Parent

What sets this book apart from other books is its approach. It is not just a math book, but a system of teaching math. Each daily lesson contains four key parts: **Review Exercises**, **Helpful Hints**, **New Material**, and **Problem Solving**. Teachers have flexibility in introducing new topics, but the book provides them with the necessary structure and guidance. The teacher can rest assured that essential math skills in this book are being systematically learned.

This easy-to-follow program requires only fifteen or twenty minutes of instruction per day. Each lesson is concise and self-contained. The daily exercises help students to not only master math skills, but also maintain and reinforce those skills through consistent review - something that is missing in most math programs. Skills learned in this book apply to all areas of the curriculum, and consistent review is built into each daily lesson. Teachers and parents will also be pleased to note that the lessons are quite easy to correct.

This book is based on a system of teaching that was developed by a math instructor over a thirty-year period. This system has produced dramatic results for students. The program quickly motivates students and creates confidence and excitement that leads naturally to success.

Please read the following "How to Use This Book" section and let this program help you to produce dramatic results with children and math students.

How to Use This Book

This book is best used on a daily basis. The first lesson should be carefully gone over with students to introduce them to the program and familiarize them with the format. It is hoped that the program will help your students to develop an enthusiasm and passion for math that will stay with them throughout their education.

As you go through these lessons every day, you will soon begin to see growth in the student's confidence, enthusiasm, and skill level. The students will maintain their mastery through the daily review.

Step 1

The Students are to complete the review exercises, showing all their work. After completing the problems, it is important for the teacher or parent to go over this section with the students to ensure understanding.

Step 2

Next comes the new material. Use the "Helpful Hints" section to help introduce the new material. Be sure to point out that it is often helpful to come back to this section as the students work independently. This section often has examples that are very helpful to the students.

Step 3

It is highly important for the teacher to work through the two sample problems with the students before they begin to work independently. Working these problems together will ensure that the students understand the topic, and prevent a lot of unnecessary frustration. The two sample problems will get the students off to a good start and will instill confidence as the students begin to work independently.

Step 4

Each lesson has problem solving as the last section of the page. It is recommended that the teacher go through this section , discussing key words and phrases, and also possible strategies. Problem solving is neglected in many math programs, and just a little work each day can produce dramatic results.

Step 5

Solutions are located in the back of the book. Teachers may correct the exercises if they wish, or have the students correct the work themselves.

Table of Contents

Whole Numbers

Addition 4

Addition with Large Numbers.. 6

Subtraction... 8

Subtraction with Zeroes.. 10

Reviewing Addition and Subtraction.. 12

Multiplying by 1-Digit Factors.. 14

Multiplying by 2-Digits.. 16

Multiplying by 3-Digits.. 18

Reviewing Multiplication... 20

Division by 1-Digit Divisors.. 22

Dividing Hundreds by 1-Digit Divisors... 24

Dividing Thousands by 1-Digit Divisors.. 26

Reviewing Division by 1-Digit Divisors.. 28

Dividing by Multiples of Ten... 30

Division by 2-Digit Divisors.. 32

Estimating when Dividing by 2-Digit Divisors.. 34

Dividing Larger Numbers with 2-Digit Divisors.. 36

Reviewing Division... 38

Integers

Addition... 42

Addition with More Than Two Integers.. 44

Subtraction... 46

Reviewing Addition and Subtraction.. 48

Multiplication.. 50

Multiplying with More than Two Factors... 52

Division... 54

Multiplication and Division - Two-Step... 56

Reviewing Multiplication and Division.. 58

Review of All Integer Operations... 60

Final Test

Final Test - Whole Numbers.. 62

Final Test - Integers.. 63

Solutions.. 64

Review Exercises

Note to the students and teachers: This section will include review problems from all the topics covered in this book. Here are some simple problems with which to get started.

1.
```
  314
+  43
```

2.
```
  603
+  24
```

3.
```
   7
   2
+  4
```

4.
```
  426
+ 313
```

5. 7 + 4 + 5 =

6.
```
  22
  16
+ 11
```

* Samples S1 and S2 may be worked together by teacher and students.

S1.
```
  243
   62
+ 514
```

S2.
```
  345
  423
+ 165
```

1.
```
  42
  56
+ 15
```

2.
```
  516
   54
+ 213
```

3.
```
  517
  326
  143
+  14
```

4.
```
  813
  236
   17
+   6
```

5.
```
  543
  227
  312
+ 415
```

6.
```
    5
   23
  416
+ 243
```

7. 415 + 436 + 317 =

8. 23 + 34 + 16 + 47 =

9. Find the sum of 443, 514, and 716.

10. Find the sum of 127, 23, 16, and 246.

1.	
2.	
3.	
4.	
5.	
6.	
7.	
8.	
9.	
10.	
Score	

Problem Solving

There are 42 students in the fifth grade, 56 students in the sixth grade, and 36 students in the seventh grade. How many students are there altogether?

Review Exercises

1. 426
    + 23

2. 526
    42
    + 163

3. Find the sum of 23, 34, and 26.

4. 342
    426
    + 325

5. 427
    316
    12
    + 423

6. 36
    324
    573
    + 11

Helpful Hints	Use what you have learned to solve each problem.

REMEMBER: 1. Line up numbers on the right side.
2. Add the ones first.
3. Regroup when necessary
4. Place commas in the answer when necessary.

* "Sum" means add.

* Samples S1 and S2 may be worked together by teacher and students.

S1. 426
    343
    + 516

S2. 16
    247
    346
    + 125

1. 702
    337
    + 225

2. 23
    45
    + 67

3. 627
    423
    504
    + 213

4. 6
    17
    289
    + 363

5. 724
    427
    395
    + 367

6. 600
    528
    396
    + 27

7. 7 + 27 + 48 + 247 =

8. 763 + 29 + 372 + 16 =

9. Find the sum of 96, 73, 44, and 75.

10. Find the sum of 213, 426, 516, and 423.

Answers:
1.
2.
3.
4.
5.
6.
7.
8.
9.
10.
Score

Problem Solving
Monique earned $52 on Monday. On Tuesday, Wednesday, and Thursday she earned $56 each day. How much did she earn altogether?

Review Exercises

1.
```
   36
   47
   16
+  92
```

2.
```
  727
  423
+  75
```

3.
```
  824
  216
  724
+ 316
```

4. 967 + 843 + 96 =

5. 72 + 16 + 49 =

6. Find the sum of 14, 15, 16, and 17.

When writing large numbers, place commas every three numerals, starting from the right. This makes them easier to read.

Example: 21 million, 234 thousand, 416

21,234,416

S1.
```
  4,236
    147
+ 3,236
```

S2.
```
   7,217
     685
+ 36,125
```

1.
```
  4,736
  2,136
+ 3,523
```

2.
```
 17,236
  3,175
+ 4,297
```

3.
```
  1,312,516
+ 2,316,475
```

4.
```
  7,137,236
    353,246
+     7,506
```

5.
```
  23,726
  17,532
   6,573
+ 23,247
```

6.
```
  72,506
   1,343
  15,208
+  5,123
```

7. Find the sum of 1,342, 1,793, and 4,562.

8. 78,623 + 4,579 + 22,396 =

9. 72,214 + 16,738 + 29,143 + 17 =

10. 103,246 + 724,516 + 72,173 =

1.
2.
3.
4.
5.
6.
7.
8.
9.
10.
Score

Problem Solving

One country has a population of 7,143,600. Another country has a population of 16,450,395. A third country has a population of 12,690,000. What is the total population of all three countries?

Review Exercises

1. 376
    493
    + 16

2. 23,463 + 7,764 + 21,976 =

3. 72,176
    43,267
    + 19,237

5. 95,623
    7,429
    19,234
    + 89,342

4. 27,967,204
    + 5,137,264

6. Find the sum of 3,426,497 and 7,136,095.

Helpful Hints	Use what you have learned to add the following problems. Practice reading your answers.

Example: 23 million, 106 thousand, 749

23,106,749

S1. 76,142
     7,913
     + 16,408

S2. 313,426
     2,462,417
     + 3,526,008

1. 27,167
    35,225
    + 16,342

2. 1,234,617
    3,426,124
    + 4,316,243

3. 7,163
    2,476
    + 3,276

4. 17,226
    26,319
    15,314
    + 27,976

5. 16
    326
    7,764
    + 12,316

6. 76,174
    133,367
    243,776
    + 712,777

7. 76,724 + 55,726 + 5,723 + 74,126 =

8. 7,134,243 + 3,712,050 + 3,516,312 =

9. Find the sum of 42,916, 47,993, and 6,716.

10. 7 + 17 + 278 + 42,563 =

1.
2.
3.
4.
5.
6.
7.
8.
9.
10.
Score

Problem Solving	A school district has three high schools. One school has 2,463 students, another school has 3,985 students, and the third school has 1,596 students. What is the total number of high school students in the district.

Review Exercises

1.
```
   33
   45
+  76
```

2.
```
    7,716
   12,727
+  23,496
```

3.
```
   7,124,563
   5,236,907
+  7,369,704
```

4.
```
   776
   397
+  442
```

5. $753,476 + 569,736 + 7,842 + 39,673 =$

6. Find the sum of 72, 96, 72, 9, and 88.

Helpful Hints	1. Line up the numbers on the right side. 2. Subtract the ones first. 3. Regroup when necessary. 4. It may be necessary to regroup more than once. 5. "Find the difference" and "how much more" means to subtract.	**Examples:**	$\begin{array}{r} 8 \\ 7\,\cancel{9}^{1}3 \\ -\ \ 75 \\ \hline 718 \end{array}$	$\begin{array}{r} 5\ 11 \\ \cancel{6}\cancel{2}^{1}3 \\ -\ 254 \\ \hline 369 \end{array}$

S1.
```
   567
-  183
```

S2.
```
   4,352
-  2,171
```

1.
```
   339
-   24
```

2.
```
   623
-   52
```

3.
```
   6,153
-   758
```

4.
```
   5,231
-  2,456
```

5.
```
   1,387
-   739
```

6.
```
   7,383
-  2,285
```

7. Find the difference between 763 and 47.

8. Subtract 527 from 3,916.

9. $7,249 - 779 =$

10. 789 is how much more than 298?

1. _____
2. _____
3. _____
4. _____
5. _____
6. _____
7. _____
8. _____
9. _____
10. _____

Score _____

Problem Solving

846 students attend Vargas School and 943 students attend Hoover School. How many more students attend Hoover School than Vargas School?

Review Exercises

1. 726
    849
    + 308

2. 721
    - 443

3. 7,614
    - 1,557

4. Find the sum of 455, 673, and 998.

5. 5,613
    - 752

6. 32,456,175
    3,214,915
    + 6,318,425

| **Helpful Hints** | 1. Use what you have learned to solve the following problems.
 2. It may be necessary to regroup more than once. | "find the difference"
 "is how much more"
 "is how much less" | All Mean Subtraction |

S1. 856
     - 297

S2. 9,613
     - 2,247

1. 715
    - 242

2. 324
    - 149

3. 3,137
    - 1,252

4. 6,334
    - 2,175

5. 7,123
    - 2,456

6. 7,156
    - 877

7. 27,346 - 15,472 =

8. Subtract 797 from 2,314.

9. 127 is how much less than 2,496?

10. Find the difference between 7,692 and 4,764.

1.	
2.	
3.	
4.	
5.	
6.	
7.	
8.	
9.	
10.	
Score	

Problem Solving Ahmir earned 86 dollars on Friday. He earned 28 dollars less than this amount on Saturday. How much did he earn on Saturday?

Review Exercises

1.
```
   395
   428
+  376
```

3. 5,123 - 1,672 =

5.
```
   376
    19
   426
+  442
```

2. 7,162 - 396 =

4. Find the difference between 8,961 and 279.

6. 3,152 - 496 =

Helpful Hints	1. Line up the numbers on the right side. 2. Subtract the ones first. 3. It may be necessary to regroup more than once.	**Examples:**

$$\begin{array}{r} \overset{6\ \ 10\ 9}{\cancel{7},\cancel{1}\cancel{0}3} \\ -\ \ \ 677 \\ \hline 6,426 \end{array} \qquad \begin{array}{r} \overset{5\ \ 9\ 9}{\cancel{6},\cancel{0}\cancel{0}0} \\ -\ 1,634 \\ \hline 4,366 \end{array}$$

S1.
```
   601
-  356
```

S2.
```
   700
-  267
```

1.
```
    90
-   67
```

2.
```
   705
-  176
```

3.
```
   4,012
-  1,345
```

4.
```
   700
-  238
```

5.
```
   8,000
-    758
```

6.
```
   5,207
-  1,539
```

7. Find the difference between 13,012 and 10,796.

8. Subtract 7,689 from 9,026.

9. 70,023 - 63,345 =

10. What number is 5,021 less than 12,506?

1.	
2.	
3.	
4.	
5.	
6.	
7.	
8.	
9.	
10.	
Score	

Problem Solving	A theatre has 1,150 seats. If 859 of them are taken, how many seats are empty?

Review Exercises

1. 500
    - 276

2. 3,196
      742
    + 276

3. Find the sum of 72, 95, 63, and 47.

4. 5,012
    - 763

5. 7,126
    - 3,147

6. 5,000 - 768 =

| **Helpful Hints** | Use what you have learned to solve the following problems. | **Examples:** | $\overset{4\ \ 11\ 9}{5,\cancel{2}\cancel{0}\cancel{1}}$ $\underline{-\ \ \ 475}$ $4,726$ | $\overset{2\ \ 9\ 9}{\cancel{3},\cancel{0}\cancel{0}\cancel{0}}$ $\underline{-\ 1,235}$ $1,765$ |

					1.

S1. 6,002
    - 1,456

S2. 7,000
    - 967

1. 900
    - 77

2. 5,004
    - 2,576

	2.

3. 2,010
    - 1,346

4. 5,600
    - 797

5. 3,102
    - 1,453

6. 50,000
    - 7,542

	3.
	4.
	5.
	6.

7. 71,005 - 2,709 =

8. Subtract 2,511 from 7,005.

9. What number is 57 less than 500?

10. 76,500 - 29,308 =

	7.
	8.
	9.
	10.
	Score

Problem Solving

A company earned 95,000 dollars this year. Last year the company earned 78,500 dollars. How much more did the company earn this year than last year?

Review Exercises

1. 375
    427
    + 28

2. 32,176
    3,724
    + 15,756

3. Find the sum of 24, 296, 752, and 897.

4. 7,121
    - 2,430

5. 7,001
    - 2,765

6. 4,000
    - 2,173

Helpful Hints	Use what you have learned to solve the following problems.

S1. 562
     7,124
     15,765
     + 1,456

S2. 4,001
     - 1,327

1. 516
    39
    + 77

2. 714
    - 253

3. 5,317
    967
    + 6,765

4. 7,000
    - 4,789

5. 6,102
    - 1,799

6. 29
    37
    46
    + 53

7. 23,076 - 13,097 =

8. Find the sum of 197, 368, and 427.

9. How much more is 921 than 143?

10. 72,176 + 7,996 + 72,199 =

1.
2.
3.
4.
5.
6.
7.
8.
9.
10.
Score

Problem Solving	Pedro wants to buy a bike that costs $850. If he has saved $755, how much more does he need to be able to buy the bike?

Review Exercises

1. 70
   + 26

2. 76,175 - 2,963 =

3. 325 + 72 + 765 + 89 =

4. 72,105
   - 7,367

5. Find the difference of
    7,723 and 10,197.

6. 72,172
    5,347
       56
   + 2,347

Helpful Hints

1. Use what you have learned to solve the following problems.
2. Line up numbers on the right.
3. Be careful when regrouping.

S1. 7,000
    - 1,345

S2. 37,673
     7,742
    + 7,396

1. 800
   - 124

2. 7,001
   - 1,237

3.   27
     347
   2,436
  + 7,998

4. 35,021
   - 7,130

5. 32 + 96 +
    33 + 37 =

6.   27
     99
     97
     68
   + 53

7. 976 + 279 + 342 + 196 =

8. 70,001 - 2,617 =

9. How much more is 7,526 than 3,017?

10. Find the sum of 22,463, 7,296, and 7,287.

1.	
2.	
3.	
4.	
5.	
6.	
7.	
8.	
9.	
10.	
Score	

Problem Solving

71,503 people visited the museum on Monday. 80,106 people visited the museum on Friday. How many more people visited the museum on Friday than on Monday?

Review Exercises

1. 501
   - 267

2. 334
   79
   617
   + 22

3. 6,002 - 3,667 =

6. 27,763
   4,245
   7,342
   +   767

4. 27 + 33 + 36 + 52 =

5. Find the difference of 8,012 and 796.

Helpful Hints	1. Line up numbers on the right. 2. Multiply the ones first. 3. Regroup when necessary. 4. "Product" means to multiply.	**Examples:** $\overset{1\ 1}{644}$ x 3 —— 1,932	$6,\overset{3\ 2}{0}76$ x 4 —— 24,304

S1. 526
    x   3

S2. 3,254
    x   6

1. 67
   x 3

2. 74
   x 6

3. 427
   x   6

4. 4,214
   x   6

5. 3,056
   x   8

6. 7,256
   x   6

7. 7,036 x 7 =

8. 4 x 3,849 =

9. Find the product of 8,762 and 6.

10. Multiply 9 and 3,872.

1.
2.
3.
4.
5.
6.
7.
8.
9.
10.
Score

Problem Solving If there are 365 days in each year, how many days are there in 7 years?

Review Exercises

1.   343
   x   4

2.   3,064
   x   7

3.   3,601 - 798 =

4.   6,000
   - 1,235

5.   375
      429
       63
   +   7

6.   763 x 4 =

Helpful Hints	Use what you have learned to solve the following problems.	*Remember Put commas in the answer if necessary and practice reading the answers.

S1.   7,234
   x   6

S2.   9,007
   x   8

1.   728
   x   6

2.   9,012
   x   9

3.   23,726
   x   7

4.   7,632
   x   6

5.   9,730
   x   9

6.   7,009
   x   4

7.   Find the product of 8 and 2,643.

8.   8,012 x 7 =

9.   6 x 23,728 =

10.   12,429 x 6 =

1.

2.

3.

4.

5.

6.

7.

8.

9.

10.

Score

Problem Solving	Ty needs a total of 450 points to win a price. If she has already earned 298 points, how many more points does she need to win the prize?

Review Exercises

1. 725
   x   6

2. 3,012
   x   9

3. 727 + 33 + 526 + 724 =

6. 375
   47
   462
   + 578

4. 23,102 is how much more than 7,256?

5. 7,001
   - 3,674

Helpful Hints

1. Line up numbers on the right.
2. Multiply the ones first.
3. Multiply the tens second.
4. Add the two products.
5. Place commas in the product if necessary.

Examples:

     43
   x 32
     86
  + 1290
   1,376

     437
   x   26
    2622
  + 8740
   11,362

S1. 43
    x 24

S2. 246
    x 53

1. 82
   x 53

2. 46
   x 17

3. 85
   x 47

4. 436
   x 25

5. 336
   x 50

6. 706
   x 47

7. Find the product of 16 and 37.

8. 92 x 47 =

9. 763 x 45 =

10. 460 x 33 =

1.
2.
3.
4.
5.
6.
7.
8.
9.
10.

Problem Solving

A school has 25 classrooms. If each classroom has 35 desks, how many desks are there altogether?

Score

Review Exercises

1. 3,015
   - 176

2. 3,145
   x   7

3.   48
   x 36

4.   423
   x   25

5. 7,163
   5,427
   + 3,136

6. 7,562 - 1,999 =

Helpful Hints

Use what you have learned to solve the following problems.
* Remember to put commas in your answer.

S1.   402
   x   26

S2.   356
   x   44

1.   26
   x 50

2.   56
   x 23

1.	
2.	
3.	
4.	
5.	
6.	
7.	
8.	
9.	
10.	
Score	

3.   55
   x 46

4.   216
   x   64

5.   248
   x   76

6.   476
   x   75

7. 92 x 103 =

8. 468 x 26 =

9. Find the product of 65 and 70.

10. Multiply 608 and 73.

Problem Solving

3,216 students attend Washington Middle School. If 2,016 of the students are boys, how many girls attend the school?

Review Exercises

1. 42
    x 36

2. 407
    x 28

3. 400
    x 35

4. 7,612
    - 1,357

5. 36 + 37 + 19 + 62 =

6. 7,000 - 2,836 =

Helpful Hints

1. Line up numbers on the right.
2. Multiply the ones first.
3. Multiply the tens second.
4. Multiply the hundreds last.
5. Add the products.
6. Put commas in the product.

Examples:

```
    243              673
x   336          x   307
   1458             4711
   9720             0000
+ 72900          + 201900
  84,078           206,611
```

S1. 153
     x 423

S2. 724
     x 526

1. 247
    x 315

2. 246
    x 137

3. 244
    x 302

4. 364
    x 503

5. 543
    x 414

6. 269
    x 400

7. 521 x 337 =

8. Find the product of 208 and 326.

9. Multiply 500 and 822.

10. 444 x 366 =

1.	
2.	
3.	
4.	
5.	
6.	
7.	
8.	
9.	
10.	
Score	

Problem Solving

A factory can produce 3,050 cars per week.
How many cars can the factory produce in one year?
(Hint: There are 52 weeks in a year.)

Review Exercises

1. 7,632
    558
    3,627
    +   36

2. 3,000
    - 2,176

3. 2,172
    x   6

4.   46
    x 23

5.   263
    x 30

6.   526
    x 704

Helpful Hints

Use what you have learned to solve the following problems.
1. Put commas in the answers.
2. Practice reading the answers.

S1.   444
    x 225

S2.   350
    x 906

1.   906
    x 717

2.   500
    x 743

3.   323
    x 435

4.   648
    x 755

5.   672
    x 986

6.   246
    x 932

7. 219 x 406 =

8. 700 x 610 =

9. 763 x 908 =

10. Multiply 723 and 847.

1.

2.

3.

4.

5.

6.

7.

8.

9.

10.

Score

Problem Solving

Each bus can hold 112 students.
How many students can 7 buses hold?

Review Exercises

1. 408
    x   6

2. Find the product
    of 24 and 36.

3. 7,665
    8,327
    + 9,342

4. 3,121
    - 2,344

5. 742
    x   6

6. 408
    x   27

Helpful Hints	Use what you have learned to solve the following problems.

S1. 426
     x   25

S2. 613
     x   425

1.   37
     x   5

2.   705
     x   6

3. 2,346
    x   7

4.   58
     x 72

5.   245
     x   63

6.   128
     x 547

7. Find the product of 15 and 726.

8. 700 x 657 =

9. 33 x 610 =

10. 308 x 407 =

1.	
2.	
3.	
4.	
5.	
6.	
7.	
8.	
9.	
10.	
Score	

Problem Solving	If there are 1,440 minutes in a day, how many minutes are there in a week?

Review Exercises

1. 365 + 19 + 342 =

2. 627
   x   5

3. 7,136
   -    807

4. 214
   x   7

5. 206
   x   77

6. 364
   x 500

| **Helpful Hints** | 1. Use what you have learned to solve the following problems.
2. Carefully line up the numbers.
3. Put commas in answers when necessary
4. Practice reading the answers. |

S1. 6,007
    x   5

S2. 347
    x   55

1. 404
   x   5

2. 7,002
   x   6

3. 8,916
   x   7

4. 96
   x 80

5. 718
   x   63

6. 424
   x 726

7. 3 x 720 =

8. 9 x 7,856 =

9. 67 x 715 =

10. 208 x 763 =

1.

2.

3.

4.

5.

6.

7.

8.

9.

10.

Score

| **Problem Solving** | 726 people attend the theatre on Saturday. On Sunday 827 people attended the theatre. How many more people attended the theatre on Sunday than on Saturday? |

Review Exercises

1. 624
    - 317

2. 347
    x   5

3. 72,196
    + 16,429

4. Find the difference between 7,912 and 995.

5. 136
    x   22

6. 200 x 309 =

Helpful Hints

1. Divide
2. Multiply
3. Subtract
4. Begin Again

Examples:

$$\begin{array}{r} 15\,r2 \\ 3\overline{)47} \\ -3\downarrow \\ \hline 17 \\ -15 \\ \hline 2 \end{array}$$

$$\begin{array}{r} 9\,r5 \\ 6\overline{)59} \\ -54 \\ \hline 5 \end{array}$$

Remember!
The remainder must be less than the divisor.

S1. $4\overline{)17}$

S2. $5\overline{)49}$

1. $3\overline{)74}$

2. $8\overline{)47}$

3. $6\overline{)78}$

4. $5\overline{)93}$

5. $7\overline{)84}$

6. $5\overline{)79}$

7. 77 ÷ 8 =

8. 96 ÷ 4 =

9. $\dfrac{65}{4}$

10. $\dfrac{47}{2}$

1.	
2.	
3.	
4.	
5.	
6.	
7.	
8.	
9.	
10.	
Score	

Problem Solving

Pencils come in boxes of 72. If the pencils are divided equally among 4 students, how many pencils will each student receive?

Review Exercises

1. $2\overline{)17}$ 2. $5\overline{)89}$ 3. $7\overline{)45}$

4. $205 - 99 =$ 5. $\begin{array}{r} 165 \\ \times\ \ 7 \\ \hline \end{array}$ 6. $\begin{array}{r} 47 \\ \times\ 33 \\ \hline \end{array}$

Helpful Hints

Use what you have learned to solve the following problems.

1. Divide   2. Multiply   3. Subtract

4. Begin Again   5. REMEMBER! The remainder must be less than the divisor.

S1. $4\overline{)73}$ S2. $6\overline{)27}$ 1. $3\overline{)29}$ 2. $3\overline{)45}$

3. $8\overline{)93}$ 4. $2\overline{)97}$ 5. $7\overline{)60}$ 6. $5\overline{)29}$

7. $77 \div 5 =$ 8. $\dfrac{99}{4}$

9. $33 \div 4 =$ 10. $65 \div 5 =$

1.
2.
3.
4.
5.
6.
7.
8.
9.
10.
Score

Problem Solving

Mr. Johnson's salary is $6,500 per month.
How much does he earn per year?
(Hint: How many months are in a year?)

23

Review Exercises

1. $6\overline{)38}$

2. $5\overline{)69}$

3. $7{,}103 - 2{,}617 =$

4. $526 \times 7 =$

5. $\begin{array}{r} 164 \\ \times\ \ 23 \\ \hline \end{array}$

6. $73 + 16 + 57 + 76 =$

Helpful Hints

1. Divide
2. Multiply
3. Subtract
4. Begin Again

Examples:

$$\begin{array}{r} 171\ r2 \\ 3\overline{)515} \\ -3\downarrow \\ \hline 21 \\ -21\downarrow \\ \hline 05 \\ -\ 3 \\ \hline 2 \end{array}$$

$$\begin{array}{r} 203 \\ 4\overline{)812} \\ -8\downarrow \\ \hline 01 \\ -\ 0\downarrow \\ \hline 12 \\ -12 \\ \hline 0 \end{array}$$

REMEMBER!
The remainder must be less than the divisor.

S1. $2\overline{)523}$

S2. $6\overline{)274}$

1. $3\overline{)972}$

2. $2\overline{)419}$

3. $7\overline{)933}$

4. $6\overline{)727}$

5. $8\overline{)916}$

6. $6\overline{)850}$

7. $3\overline{)936}$

8. $5\overline{)607}$

9. $3\overline{)776}$

10. $8\overline{)717}$

1.
2.
3.
4.
5.
6.
7.
8.
9.
10.
Score

Problem Solving

A bakery produced 324 cookies. If 9 cookies are placed into each package, how many packages will the bakery need?

Review Exercises

1. 3$\overline{)57}$ 2. 7$\overline{)89}$ 3. 5$\overline{)616}$

4. 5$\overline{)374}$ 5. 7$\overline{)924}$ 6. 7$\overline{)450}$

Helpful Hints	Use what you have learned to solve the following problems.

S1. 2$\overline{)375}$ S2. 4$\overline{)377}$ 1. 8$\overline{)916}$ 2. 8$\overline{)630}$

3. 7$\overline{)679}$ 4. 8$\overline{)976}$ 5. 4$\overline{)508}$ 6. 6$\overline{)248}$

7. 3$\overline{)400}$ 8. 7$\overline{)693}$ 9. 5$\overline{)778}$ 10. 6$\overline{)609}$

1.
2.
3.
4.
5.
6.
7.
8.
9.
10.
Score

Problem Solving	The Martinez family took a vacation and drove 450 miles each day for 12 days. How many miles did they drive during their vacation?

Review Exercises

1. $3\overline{)69}$

3.   213
     x   47

5. 900 x 614 =

2. $5\overline{)667}$

4.   7,710
     x   697

6.   712
         14
     +   47

Helpful Hints	1. Divide 2. Multiply 3. Subtract 4. Begin Again

Examples:

```
      1708
   3)5124
    - 3
      21
    - 21
      02
    - 0
      24
    - 24
       0
```

```
       448 r2
    4)1794
    - 16
      19
    - 16
      34
    - 32
       2
```

REMEMBER! The remainder must be less than the divisor.

S1. $3\overline{)9052}$ S2. $5\overline{)3352}$ 1. $2\overline{)9235}$ 2. $2\overline{)7362}$

3. $6\overline{)6821}$ 4. $5\overline{)2240}$ 5. $5\overline{)7666}$ 6. $4\overline{)7123}$

7. $5\overline{)61,235}$ 8. $2\overline{)24,362}$ 9. $4\overline{)41,236}$ 10. $5\overline{)22,014}$

1.

2.

3.

4.

5.

6.

7.

8.

9.

10.

Problem Solving	5 students opened a business and earned $16,550. If they divided the money equally, how much would each student receive?	Score

Review Exercises

1. $3\overline{)526}$

2. $3\overline{)605}$

3. $7\overline{)2634}$

4. $5\overline{)6175}$

5. $6\overline{)13,252}$

6. $7,100 - 768 =$

Helpful Hints

1. Use what you have learned to solve the following problems.
2. Remainders must be less than the divisor.
3. Zeroes may sometimes appear in the quotient.

S1. $6\overline{)3976}$ S2. $5\overline{)13,976}$ 1. $3\overline{)4123}$ 2. $5\overline{)1726}$

3. $8\overline{)8902}$ 4. $6\overline{)5324}$ 5. $6\overline{)9309}$ 6. $8\overline{)8016}$

7. $3\overline{)16,723}$ 8. $7\overline{)22,305}$ 9. $4\overline{)61,053}$ 10. $8\overline{)37,716}$

1.	
2.	
3.	
4.	
5.	
6.	
7.	
8.	
9.	
10.	
Score	

Problem Solving

Yana took a 3-day hiking trip. The first day she hiked 13 miles, the second day she hiked 14 miles, and the third day she hiked 17 miles. How many miles did she hike altogether during the trip?

Review Exercises

1. 337
    96
    + 349

2. 32,105
    - 6,737

3. 427
    x   26

4. 4⟌500

5. 7⟌1356

6. Find the sum of 39,764 and 79,743.

| Helpful Hints | 1. Divide 2. Multiply 3. Subtract 4. Begin again | * Remainders must be less than the divisor. * Zeroes may sometimes appear in the quotient. |

S1. 3⟌901   S2. 7⟌7563   1. 5⟌402   2. 5⟌8500

3. 4⟌8413   4. 8⟌3566   5. 6⟌1526   6. 5⟌6000

7. 4⟌8009   8. 9⟌8765   9. 7⟌6329   10. 4⟌3217

1.
2.
3.
4.
5.
6.
7.
8.
9.
10.
Score

| Problem Solving | A ream of paper contains 500 sheets. How many sheets of paper are there in 25 reams? |

Review Exercises

1. Find the difference between 7,026 and 4,567.

2. $$637 \times 502$$

3. $7\overline{)600}$

4. $5\overline{)5003}$

5. $2\overline{)3051}$

6. $7\overline{)1969}$

Helpful Hints

Use what you have learned to solve the following problems.

S1. $3\overline{)369}$

S2. $6\overline{)8772}$

1. $5\overline{)307}$

2. $9\overline{)157}$

3. $6\overline{)1213}$

4. $7\overline{)7012}$

5. $6\overline{)9736}$

6. $4\overline{)1398}$

7. $8\overline{)13,423}$

8. $5\overline{)14,387}$

9. $6\overline{)71,234}$

10. $5\overline{)15,355}$

1.	
2.	
3.	
4.	
5.	
6.	
7.	
8.	
9.	
10.	
Score	

Problem Solving

A dairy produced 4,448 gallons of milk. If the milk is put into containers that hold 8 gallons each, how many containers will be needed?

Review Exercises

1. $6\overline{)718}$ 2. $2{,}007 - 865 =$ 3. $\begin{array}{r} 324 \\ \times\ 700 \end{array}$

4. $5\overline{)5007}$ 5. $6\overline{)1399}$ 6. $\begin{array}{r} 3{,}966 \\ 7{,}723 \\ +\ 4{,}564 \end{array}$

Helpful Hints

1. Divide
2. Multiply
3. Subtract
4. Begin again

Examples:

$$\begin{array}{r} 12\ \text{r}49 \\ 60\overline{)769} \\ -60\downarrow \\ \hline 169 \\ -120 \\ \hline 49 \end{array} \qquad \begin{array}{r} 44\ \text{r}5 \\ 40\overline{)1765} \\ -160\downarrow \\ \hline 165 \\ -160 \\ \hline 5 \end{array}$$

S1. $30\overline{)176}$ S2. $40\overline{)5732}$ 1. $70\overline{)238}$ 2. $50\overline{)376}$

3. $50\overline{)438}$ 4. $70\overline{)829}$ 5. $30\overline{)1682}$ 6. $20\overline{)1396}$

7. $40\overline{)8972}$ 8. $90\overline{)9096}$ 9. $40\overline{)3786}$ 10. $40\overline{)2396}$

1.
2.
3.
4.
5.
6.
7.
8.
9.
10.

Score

Problem Solving

A store puts eggs into boxes of a dozen. If there are 3,072 eggs, how many boxes will be needed?

Review Exercises

1. $2\overline{)69}$

2. $7\overline{)7019}$

3. $367 + 246 +$
   $721 + 243 =$

4. $36 \times 304 =$

5. $6,108 - 976 =$

6. $3\overline{)6123}$

Helpful Hints

1. Use what you have learned to solve the following problems.
2. Zeroes may sometimes appear in the quotient.
3. Remember that remainders must be less than the divisor.

S1. $50\overline{)137}$

S2. $60\overline{)7612}$

1. $20\overline{)396}$

2. $20\overline{)187}$

3. $30\overline{)2312}$

4. $30\overline{)7396}$

5. $70\overline{)5724}$

6. $70\overline{)9988}$

7. $50\overline{)9976}$

8. $40\overline{)1766}$

9. $80\overline{)2209}$

10. $50\overline{)12,076}$

1.
2.
3.
4.
5.
6.
7.
8.
9.
10.
Score

Problem Solving

A school has 24 classes. Each class contains 32 students. How many students are there altogether in the school?

Review Exercises

1.   463
        87
   + 496

2.   7,015
   - 1,247

3.   963 - 713 =

4. 60 ⟌ 729

5. 70 ⟌ 367

6. 40 ⟌ 5276

Helpful Hints	Sometimes it is easier to mentally round the divisor to the nearest power of ten.

Examples:
$$22 \overline{)715} \quad \begin{array}{c} 32\ r11 \\ \hline \end{array}$$
   - 66↓
      55
   - 44
      11

Think of:
20 ⟌ 715

S1. 32 ⟌ 673

S2. 32 ⟌ 279

1. 41 ⟌ 936

2. 28 ⟌ 617

3. 58 ⟌ 697

4. 83 ⟌ 762

5. 34 ⟌ 826

6. 31 ⟌ 976

7. 62 ⟌ 759

8. 53 ⟌ 609

9. 21 ⟌ 936

10. 42 ⟌ 866

1.

2.

3.

4.

5.

6.

7.

8.

9.

10.

Score

Problem Solving	A car traveled 448 miles. For each 32 miles it traveled, it consumed one gallon of gas. How many gallons of gas did it consume in traveling 448 miles?

Review Exercises

1. 326
   x 35

2. 324
   x 500

3. 27 + 89 + 76 + 14 =

5. 50$\overline{)826}$

6. 90$\overline{)7372}$

4. 3,000 - 756 =

Helpful Hints

Use what you have learned to solve the following problems.

*Remember: Sometimes mentally rounding the divisor helps to make the problem easier.

Examples:
$$27\overline{)998}$$
$$\begin{array}{r} 36 \\ \hline 998 \\ -81\downarrow \\ \hline 188 \\ -162 \\ \hline 26 \end{array}$$

Think of:
$$30\overline{)998}$$

S1. 41$\overline{)912}$

S2. 51$\overline{)367}$

1. 52$\overline{)706}$

2. 39$\overline{)876}$

3. 47$\overline{)613}$

4. 22$\overline{)796}$

5. 12$\overline{)393}$

6. 25$\overline{)618}$

7. 49$\overline{)536}$

8. 22$\overline{)684}$

9. 28$\overline{)976}$

10. 42$\overline{)750}$

1. _____
2. _____
3. _____
4. _____
5. _____
6. _____
7. _____
8. _____
9. _____
10. _____

Score

Problem Solving

A theater had 12 rows of seats with 16 seats in a row.
If only 7 seats were empty, how many seats were occupied?

Review Exercises

1. 30⟌98   2. 60⟌793   3. 60⟌483

4. 32⟌426   5. 32⟌275   6. 28⟌405

| **Helpful Hints** | Sometimes it is necessary to correct your estimate. | Examples: | $\begin{array}{r}6\\63\overline{)374}\\-378\end{array}$ - too large | $\begin{array}{r}5\text{r}60\\63\overline{)374}\\-315\\\hline60\end{array}$ |

S1. 74⟌293   S2. 43⟌821   1. 38⟌197   2. 88⟌522

1.
2.
3.
4.
5.
6.
7.
8.
9.
10.
Score

3. 18⟌987   4. 22⟌178   5. 32⟌163   6. 14⟌886

7. 34⟌649   8. 42⟌829   9. 36⟌721   10. 18⟌778

| **Problem Solving** | A machine can produce 75 parts in one hour. How many parts can it produce in 12 hours? |

Review Exercises

1. 26
    x 3

2. 2,476
    x   7

3. 64
    x 23

4. 207
    x 25

5. 627
    x 400

6. 763
    x 509

Helpful Hints

Use what you have learned to solve the following problems.
*Remember: Sometimes it is necessary to correct your estimate.

S1. 29$\overline{)862}$ S2. 22$\overline{)830}$ 1. 27$\overline{)851}$ 2. 25$\overline{)805}$

3. 19$\overline{)572}$ 4. 23$\overline{)476}$ 5. 31$\overline{)927}$ 6. 12$\overline{)588}$

7. 18$\overline{)787}$ 8. 42$\overline{)829}$ 9. 14$\overline{)886}$ 10. 21$\overline{)178}$

1.

2.

3.

4.

5.

6.

7.

8.

9.

10.

Score

Problem Solving

Children's tickets to a show are 5 dollars each and adult tickets are 7 dollars each. What would the total price be for 5 children's tickets and 4 adult tickets?

35

Review Exercises

1.
```
   721
-  356
```

2. 755 - 288 =

3.
```
   60,123
-  29,465
```

4.
```
   779
   643
   247
+   96
```

5. 637 + 975 + 734 + 623 =

6. 7,256 is how much more than 5,909?

Helpful Hints

Use what you have learned to solve the following problems.

*Remember: Mentally round your divisor to the nearest power of ten.

Example:
```
        216 r20
   32 ) 6932
      - 64▼
        53
      - 32▼
        212
      - 192
         20
```

Think of:
```
   30 ) 6932
```

S1. 43) 9250   S2. 32) 6635   1. 21) 2645   2. 51) 7563

3. 42) 9005   4. 31) 5126   5. 27) 8056   6. 25) 7651

7. 81) 2654   8. 22) 1765   9. 28) 3976   10. 12) 9006

1.	
2.	
3.	
4.	
5.	
6.	
7.	
8.	
9.	
10.	
Score	

Problem Solving

A car traveled at the speed of 55 miles per hour for 8 hours. What was the total distance traveled?

Review Exercises

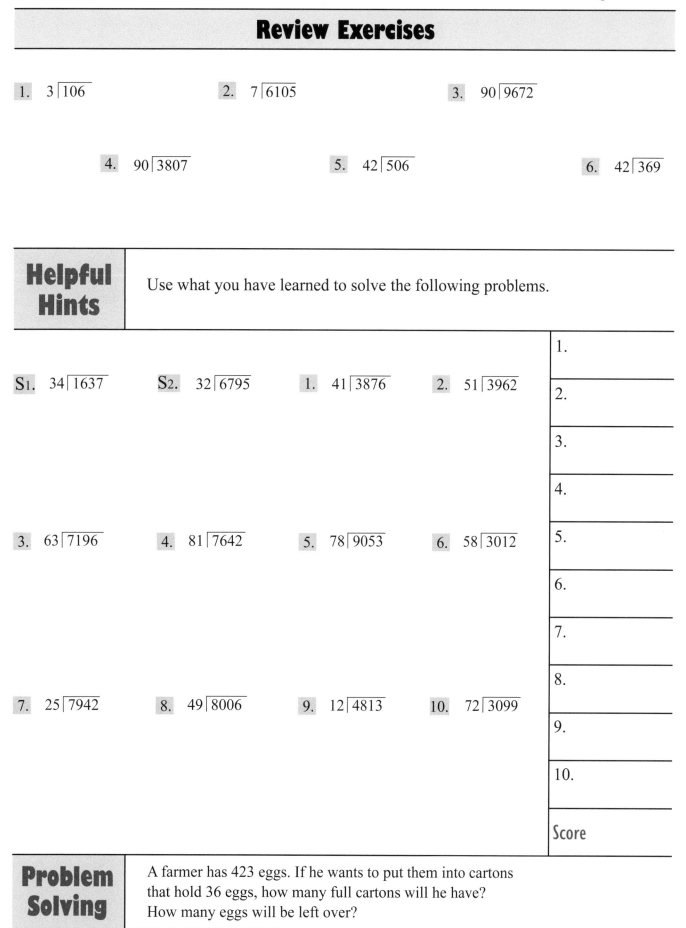

1. 3⟌106   2. 7⟌6105   3. 90⟌9672

4. 90⟌3807   5. 42⟌506   6. 42⟌369

Helpful Hints

Use what you have learned to solve the following problems.

	1.
S1. 34⟌1637 S2. 32⟌6795 1. 41⟌3876 2. 51⟌3962	2.
	3.
	4.
3. 63⟌7196 4. 81⟌7642 5. 78⟌9053 6. 58⟌3012	5.
	6.
	7.
	8.
7. 25⟌7942 8. 49⟌8006 9. 12⟌4813 10. 72⟌3099	9.
	10.
	Score

Problem Solving

A farmer has 423 eggs. If he wants to put them into cartons that hold 36 eggs, how many full cartons will he have? How many eggs will be left over?

Review Exercises

1. 72,096
    7,423
    + 27,564

2. 7,505
    - 3,176

3. Find the sum of
   79,216 and 76,719.

4. Find the product of
   725 and 301.

5. Find the difference of
   6,000 and 890.

6.   336
   x   22

Helpful Hints	Use what you have learned to solve the following problems.	*Mentally round 2-digit divisors when helpful.
		*Remainders must be less than the divisor.

S1. 81⟌7716	S2. 28⟌1561	1. 2⟌77	2. 6⟌1989	1.
				2.
				3.
				4.
3. 5⟌1397	4. 40⟌647	5. 90⟌706	6. 90⟌3762	5.
				6.
				7.
				8.
7. 38⟌966	8. 23⟌278	9. 61⟌3344	10. 32⟌8096	9.
				10.
				Score

Problem Solving

There were 96 students on a bus. At one stop 28 students got off. At the next stop 17 students got off. How many students were left on the bus?

Review Exercises

1. $2\overline{)68}$

2. $3\overline{)1964}$

3. $30\overline{)279}$

4. $30\overline{)1367}$

5. $62\overline{)707}$

6. $62\overline{)4537}$

Helpful Hints

Use what you have learned to solve the following problems.

S1. $61\overline{)5872}$ S2. $27\overline{)2096}$ 1. $5\overline{)97}$ 2. $8\overline{)2605}$

3. $9\overline{)7002}$ 4. $60\overline{)967}$ 5. $60\overline{)396}$ 6. $80\overline{)6776}$

7. $76\overline{)396}$ 8. $76\overline{)962}$ 9. $76\overline{)1796}$ 10. $76\overline{)9908}$

1.

2.

3.

4.

5.

6.

7.

8.

9.

10.

Score

Problem Solving

Susan works 40 hours per week. If she is paid 12 dollars per hour, how much does she earn in 2 weeks?

Review of All Whole Number Operations

1.
```
   456
    27
+  626
```

2.
```
   819
   746
   696
+  638
```

3. $7,964 + 895 + 72,528 =$

4. $7,096 + 2,716 + 779 + 84 =$

5. $7,009 + 868 + 19 + 578 =$

6.
```
   742
-  397
```

7.
```
  6,291
- 3,476
```

8. $9,051 - 2,766 =$

9. $8,000 - 3,988 =$

10. $9,051 - 2,766 =$

11.
```
    87
x    3
```

12.
```
  6,542
x     7
```

13.
```
    49
x   63
```

14.
```
   809
x   76
```

15.
```
   409
x  278
```

16. $3 \overline{) 425}$

17. $6 \overline{) 1697}$

18. $40 \overline{) 867}$

19. $53 \overline{) 7976}$

20. $38 \overline{) 1698}$

1.
2.
3.
4.
5.
6.
7.
8.
9.
10.
11.
12.
13.
14.
15.
16.
17.
18.
19.
20.

Final Review of All Whole Number Operations

1. 767
    455
    396
    + 228

2. 72,367
    9,768
    + 7,709

3. 16,221 + 6,872 + 9,796 + 15 =

4. 7 + 77 + 777 + 7,777 =

5. 9,697 + 7,636 + 8,964 =

6. 700
    - 265

7. 9,001
    - 786

8. 52,153 - 7,654 =

9. 5,000 - 2,357 =

10. 72,591 - 16,784 =

11. 308
     x   7

12. 7,010
     x   9

13. 87
     x 29

14. 978
     x 39

15. 963
     x 801

16. 7) 412

17. 6) 1392

18. 40) 239

19. 63) 8974

20. 29) 17,236

1.	
2.	
3.	
4.	
5.	
6.	
7.	
8.	
9.	
10.	
11.	
12.	
13.	
14.	
15.	
16.	
17.	
18.	
19.	
20.	

Review Exercises

1. 775 + 639 + 426 =

2. 7,163
    -   269

3. Find the difference between 7,455 and 3,672.

4. 7$\overline{)1637}$

5. 70$\overline{)367}$

6. 40$\overline{)719}$

Helpful Hints

-4 -3 -2 -1 0 1 2 3 4

Integers to the left of zero are negative and less than zero. Integers to the right of zero are positive and greater than zero. When two integers are on a number line, the one farthest to the right is greater.

Hint: Always find the sign of the answer first when working these problems.

Examples: The sum of two negatives is a negative.

-7 + -5 = -   (The sign is negative.)

$$\begin{array}{r} 7 \\ + 5 \\ \hline 12 \end{array} = \boxed{-12}$$

When adding a negative and a positive, the sign is the same as the integer farthest from zero, then subtract.

-7 + 9 = +   (The sign is positive.)

$$\begin{array}{r} 9 \\ - 7 \\ \hline 2 \end{array} = \boxed{+2}$$

S1. -7 + 13 =

S2. -16 + -7 =

1. -17 + 26 =

2. -14 + -4 =

3. 56 + -72 =

4. -14 + -13 =

5. -9 + 16 =

6. -47 + 85 =

7. -102 + -78 =

8. 65 + -94 =

9. -23 + -47 =

10. 65 + -85 =

1.
2.
3.
4.
5.
6.
7.
8.
9.
10.
Score

Problem Solving

There are 320 students going on a field trip. If each bus holds 55 students, how many buses will be needed for the field trip?

Review Exercises

1. -32 + 18 =

2. 17 + -12 =

3. -32 + -46 =

4.   453
     x 327

5. 32$\overline{)409}$

6. 32$\overline{)197}$

Helpful Hints	Use what you have learned to solve the following problems. *Remember to find the sign of the answer first.

S1. -96 + 105 =

S2. -99 + -86 =

1. -67 + 58 =

2. -235 + -701 =

3. -56 + 19 =

4. -95 + -46 =

5. -163 + 200 =

6. -423 + 208 =

7. -525 + -376 =

8. 924 + -1,023 =

9. -346 + -295 =

10. 650 + -496 =

1.	
2.	
3.	
4.	
5.	
6.	
7.	
8.	
9.	
10.	
Score	

Problem Solving	Maria bought a CD that cost 19 dollars. If she paid with a 50 dollar bill, how much change will she receive?

43

Review Exercises

1.   3,009
      x   7

2.   767
           96
       + 394

3.   72,052
       - 13,654

4. -75 + 60 =

5. -92 + -96 =

6. -105 + 142 =

Helpful Hints	When adding more than two integers, group the negatives and positives separately, then add.	**Examples:** $-6 + 4 + -5 =$ $\quad$ 11 $\quad$ $-11 + 4 = -$ $\quad$ - 4 (Sign is negative.) $\quad$ $\overline{7} = \boxed{-7}$	$7 + -3 + -8 + 6 =$ $\quad$ 13 $\quad$ $-11 + 13 = +$ $\quad$ - 11 (Sign is positive.) $\quad$ $\overline{2} = \boxed{+2}$

S1. -4 + 7 + -5 =

S2. -7 + 6 + -8 + 4 =

1. -7 + -5 + 15 =

2. 9 + -4 + -8 =

3. -16 + 22 + -12 =

4. 8 + -6 + 4 + 7 =

5. -22 + 40 + -24 =

6. -12 + 8 + -10 + 6 =

7. -12 + -7 + -14 =

8. -7 + 8 + -3 + -7 =

9. -40 + 18 + -16 + 12 =

10. -84 + 30 + -35 =

1.
2.
3.
4.
5.
6.
7.
8.
9.
10.
Score

Problem Solving

A student received test scores of 84, 90, and 93.
What was his average score?

Review Exercises

1. $40\overline{)967}$ 2. $40\overline{)384}$ 3. $32\overline{)697}$

4. $32\overline{)1987}$ 5. $-4 + 8 + -9 + 7 =$ 6. $-35 + 40 + -77 =$

Use what you have learned to solve the following problems.
Remember to group the negatives and positives separately, then add.

S1. $-7 + 9 + -8 =$ S2. $-5 + 12 + -13 + 15 =$

1. $-6 + -8 + 7 =$ 2. $15 + -7 + -30 =$

3. $-50 + 17 + -22 =$ 4. $16 + -12 + 18 + -8 =$

5. $-52 + 32 + -14 =$ 6. $-20 + 14 + -12 + 32 =$

7. $62 + -101 + 15 =$ 8. $-35 + -36 + -37 =$

9. $-12 + 21 + -16 + 40 =$ 10. $-92 + 58 + -23 =$

1.	
2.	
3.	
4.	
5.	
6.	
7.	
8.	
9.	
10.	
Score	

A business needs 600 postcards to mail to customers.
If postcards come in packages of 25, how many packages
does the business need to buy?

Review Exercises

1. -17 + 5 + -16 =

2. -5 + -6 + 7 + 12 =

3. -14 + 25 + -32 + 6 =

4. 36 + 72 + 14 +
   96 + 23 =

5. 2,015 - 786 =

6.   786
      x   22

Helpful Hints

* To subtract integers means to add to its opposite. **Examples:**

-3 - -8 = 8 -3 + 8 = + - 3 (Sign is positive.) 5 = (+5)	8 - 10 = 10 8 + -10 = - - 8 (Sign is negative.) -2 = (-2)	6 - -7 = 7 6 + 7 = + + 6 (Sign is positive.) 13 = (+13)	

S1. -7 - 6 =

S2. -7 - -8 =

1. 3 - -12 =

2. 16 - 19 =

3. -14 - -22 =

4. -17 - 14 =

5. 50 - -16 =

6. 48 - 14 =

7. -8 - 12 =

8. -72 - -54 =

9. -39 - 54

10. -63 - -94 =

1.
2.
3.
4.
5.
6.
7.
8.
9.
10.
Score

Problem Solving

At night the temperature was 44 degrees. By morning it had dropped 52 degrees. What was the temperature in the morning?

Review Exercises

1. -7 + 6 + -8 =

2. 16 + -22 + -10 + 6 =

3. 32 + -46 + -16 =

4. -35 - 7 =

5. 16 - -4 =

6. 32 - 48 =

Helpful Hints

Use what you have learned to solve the following problems.
Remember, to subtract means to add its opposite.

Examples: -8 - 20 = 8 + -20   -8 - -20 = -8 + 20

S1. -12 - 16 =

S2. 50 - -62 =

1. -22 - -60 =

2. 15 - 23 =

3. -24 - -36 =

4. -14 - 32 =

5. 55 - -16 =

6. -39 - 40 =

7. -6 - 5 - 3 =

8. -102 - -150 =

9. -220 - 214 =

10. -58 - -42 =

1.

2.

3.

4.

5.

6.

7.

8.

9.

10.

Score

Problem Solving

Together, Juan and James have earned 520 dollars.
If Juan has earned 325 dollars, how much has James earned?

Review Exercises

1. -7 - 9 =   2. -12 - -10 =   3. 18 - -22 =

4. 16 + -7 + 12 =   5. -8 + 6 + -7 + 5 =   6. -22 + -72 + -13 =

Helpful Hints

Use what you have learned to solve the problems on this page. **Examples:**

| -7 + 4 + -3 + 2 = 10
-10 + 6 = - - 6
(Sign is negative.) ‾‾4 = (-4) | -7 - -6 = 7
-7 + 6 = - - 6
(Sign is negative.) ‾‾1 = (-1) | 15 - 36 = 36
15 + -36 = - - 15
(Sign is negative.) ‾‾21 = (-21) |

S1. -94 + 48 =   S2. 15 - -7 =   1. -42 + -63 = | 1. |

2. -95 + 110 =   3. -9 - 12 =   4. 40 - -20 = | 2. |
                                                                              | 3. |
                                                                              | 4. |
5. -12 + 5 + -16 =   6. 20 + -7 + 4 + -8 =   7. 64 - 93 = | 5. |
                                                                              | 6. |
                                                                              | 7. |
8. 7 - -12 =   9. -425 + 501 =   10. -723 - 201 = | 8. |
                                                                              | 9. |
                                                                              | 10. |

Problem Solving

If the temperature was -12° at midnight and by 6:00 A.M.
it had dropped another 22°, what was the temperature
at 6:00 A.M.?

Score

Review Exercises

1. 336 + -521 =

2. -75 - 96 =

3. -402 + 763 =

4. 91 - -65 =

5. -256 + -758 =

6. -7 - 9 - 6 =

Helpful Hints

Use what you have learned to solve the following problems.
If you need help, refer to the examples on the previous page.

S1. -763 - 202 =

S2. 95 - -62 =

1. -29 - 36 =

2. -428 + 500 =

3. 50 - -21 =

4. 72 - 125 =

5. 65 + 12 + -52 + 16 =

6. -37 + -16 + -42 =

7. -95 - 24 =

8. -55 - -30 =

9. -33 - 35 =

10. -316 + -422 =

1.

2.

3.

4.

5.

6.

7.

8.

9.

10.

Score

Problem Solving

If the temperature was -12° at 6:00 A.M. and by noon
it had risen 33°, what was the temperature at noon?

Review Exercises

1. 337
    98
    324
    + 7

2. 1,712
    - 963

3. 304
    x 27

4. 63)1724

5. -27 - 37 =

6. 27 - -37 =

Helpful Hints

The product of two integers with different signs is negative.

The product of two integers with the same sign is positive. (• means multiply).

Examples:

$7 • -16 = -$ (Sign is negative.)

$$\begin{array}{r} 16 \\ \times\ 7 \\ \hline 112 \end{array} = \boxed{-112}$$

$-8 • -7 = +$ (Sign is positive.)

$$\begin{array}{r} 8 \\ \times\ 7 \\ \hline 56 \end{array} = \boxed{+56}$$

S1. -4 x -12 =

S2. -17 • 8 =

1. -6 • -29 =

2. 16 • -5 =

3. -36 • -14 =

4. 27 • -22 =

5. -40 • 36 =

6. 9 x -19 =

7. -8 • -7 =

8. -24 • -16 =

9. 34 x -8 =

10. -17 • -38 =

1.	
2.	
3.	
4.	
5.	
6.	
7.	
8.	
9.	
10.	
Score	

Problem Solving

Bill, Robert, and Olga together earned 520 dollars on Monday and 470 dollars on Tuesday. If they wanted to divide the money equally, how much would each person get?

Review Exercises

1. $-6 \cdot -7 =$

2. $12 \times -13 =$

3. $-15 \cdot -20 =$

4. $63 - 75 =$

5. $-66 + 95 =$

6. $-75 - -90 =$

Helpful Hints

Use what you have learned to solve the following problems.

Remember: The product of two integers with different signs is negative.
The product of two integers with the same signs is positive.

S1. $-15 \cdot 12 =$

S2. $-16 \cdot -20 =$

1. $-5 \cdot -22 =$

2. $5 \times -22 =$

3. $-36 \times -12 =$

4. $42 \times -10 =$

5. $-320 \times 5 =$

6. $-75 \cdot -5 =$

7. $6 \cdot -220 =$

8. $-18 \cdot -40 =$

9. $32 \cdot -18 =$

10. $-160 \cdot -15 =$

1.
2.
3.
4.
5.
6.
7.
8.
9.
10.
Score

Problem Solving

A rope is 525 feet long. If it is cut into 5 pieces of equal length, how long will each piece be?

Review Exercises

1. -9 + 6 + -12 + 8 =

2. -7 - -9 =

3. -6 • -7 =

4. -6 x -42 =

5. -7 + -9 + -7 =

6. 16 - -20 =

Helpful Hints

When multiplying more than two integers, group them in pairs to simplify.

An integer next to parentheses means to multiply.

Examples:

2 • -3 (-6) =
(2 • -3) (-6) =
-6 (-6) = +
(Sign is positive.)

    6
  x 6
  ___
   36 = (+36)

-2 • -3 • 4 • -2 =
(-2 • -3) • (4 • -2) =
6 • -8 = -
(Sign is negative.)

    8
  x 6
  ___
   48 = (-48)

S1. -2 • 8 • -4 =

S2. (-2)(-5) • -3 =

1. 4 (-3) • 5 =

2. -3 • -6 (-7) =

3. 3 • -4 • -5 • 6 =

4. 5 (3) • -2 x (-6) =

5. 2 • -3 • -1 • -2 =

6. (-6) (-4) (-5) =

7. -3 (-4) • 1 (-4) =

8. 4 (-6) • 2 (-4) =

9. (-6) (-7) (2) (3) =

10. 11 (-10) (-4) =

1.

2.

3.

4.

5.

6.

7.

8.

9.

10.

Score

Problem Solving

If a plane can travel 460 miles per hour, how far can it travel in 8 hours?

Review Exercises

1. -72 + 96 =

2. -53 - 52 =

3. 56 + -62 - -50 =

4. 7 • -12 =

5. -2 (-6) (-3) =

6. (-2) (-3) (-4) =

Helpful Hints

Use what you have learned to solve the following problems.
Remember, when multiplying more than two integers,
group them into pairs to simplify.

			1.
S1. -9 • 6 • -7 =	S2. -10 (-6) • (-3) =	1. 4 (-5) (-6) =	2.
			3.
			4.
2. -7 • -3 (-5) =	3. -2 • -3 • 4 • -6 =	4. 5 (6) • -5 • (-3) =	5.
			6.
5. (-5) (-8) (-5) =	6. (2) (-3) (4) (-5) =	7. (-7) (-4) • 2 (-6) =	7.
			8.
			9.
8. 5 (-6) • 4 (-3) =	9. (-6) (7) (-3) (2) =	10. -12 (3) (-6) =	10.
			Score

Problem Solving

A farmer owns 750 cows. If he decides to sell one-half
of them, how many cows will he sell?

Review Exercises

1. $-7 + 12 + -6 =$

2. $-7 - -9 =$

3. $3 + -7 - -9 =$

4. $-63 - 72 - -60 =$

5. $3 \, (-3) \, (-2) =$

6. $-5 \, (2) \, (2) \, (-3) =$

Helpful Hints	The quotient of two integers with different signs is negative. The quotient of two integers with the same sign is positive. (Hint: determine the sign, then divide.).	**Examples:** $36 \div -4 = -$ (Sign is negative.) $4\overline{\smash{)}36}$ $\begin{array}{r} 9 \\ \underline{-36} \\ 0 \end{array}$ $= \boxed{-9}$ $\dfrac{-123}{-2} = +$ (Sign is positive.) $3\overline{\smash{)}123}$ $\begin{array}{r} 41 \\ \underline{-12\downarrow} \\ 3 \end{array}$ $= \boxed{+4}$

S1. $12 \div -4 =$

S2. $\dfrac{-75}{-5}$

1. $-72 \div 4 =$

2. $336 \div -7 =$

3. $\dfrac{-105}{-5}$

4. $204 \div -4 =$

5. $\dfrac{-130}{5}$

6. $576 \div -12 =$

7. $56 \div -7 =$

8. $357 \div -21 =$

9. $\dfrac{-65}{-5}$

10. $-684 \div -36 =$

1.

2.

3.

4.

5.

6.

7.

8.

9.

10.

Score

Problem Solving	Eva is inviting 70 people to a party. She plans to provide one soft drink for each person invited. If soft drinks come in packs of six, how many packs must she buy?

Review Exercises

1. $7\overline{)23}$

2. $7\overline{)1396}$

3. $60\overline{)896}$

4. $60\overline{)394}$

5. $32\overline{)186}$

6. $32\overline{)489}$

Helpful Hints	Use what you have learned to solve the following problems. The quotient of two integers with different signs is negative. The quotient of two integers with the same sing is positive.

S1. $96 \div -3 =$

S2. $\dfrac{-90}{-15}$

1. $-72 \div 4 =$

2. $324 \div -4 =$

3. $\dfrac{-115}{5}$

4. $\dfrac{-2121}{-3}$

5. $864 \div -12 =$

6. $\dfrac{-110}{-5}$

7. $-80 \div -5 =$

8. $\dfrac{65}{-13}$

9. $-104 \div 4 =$

10. $-2001 \div -3 =$

1.

2.

3.

4.

5.

6.

7.

8.

9.

10.

Score

Problem Solving	A car traveled 295 miles in 5 hours. What was the car's average speed?

Review Exercises

1. 32 (-3) =

2. (-2) (-6) =

3. (-3) 2 (-4) =

4. 200 ÷ -4 =

5. $\dfrac{60}{-15}$

6. -90 ÷ -15 =

Helpful Hints

Use what you have learned to solve problems like these.

Examples:

$$\dfrac{-36 \div -9}{4 \div -2} = \dfrac{4}{-2} = \boxed{-2}$$
(Sign is negative.)

$$\dfrac{4 \times -8}{-8 \div 2} = \dfrac{-32}{-4} = \boxed{+8}$$
(Sign is positive.)

S1. $\dfrac{-15 \div -3}{5 \div -1} =$

S2. $\dfrac{3 \cdot (-8)}{-8 \div -4} =$

1. $\dfrac{-32 \div 4}{2 \cdot -2} =$

2. $\dfrac{-6 \times -5}{-30 \div -3} =$

3. $\dfrac{4 \cdot (-6)}{(-2)(-3)} =$

4. $\dfrac{-4 \cdot -9}{8 \div -4} =$

5. $\dfrac{-36 \div -6}{-10 \div 5} =$

6. $\dfrac{-24 \div -3}{-2 \cdot -2} =$

7. $\dfrac{75 \div -25}{3 \div -1} =$

8. $\dfrac{-42 \div -2}{14 \div -2} =$

9. $\dfrac{45 \div 5}{-9 \div 3} =$

10. $\dfrac{-56 \div -7}{-36 \div -9} =$

1.
2.
3.
4.
5.
6.
7.
8.
9.
10.
Score

Problem Solving

Kale has a library book that is seven days overdue.
For each day overdue he must pay 25 cents.
How much must he pay the library for his overdue book?

Review Exercises

1. Find the difference between 709 and 688.

2. Find the product of 42 and 604.

3. Find the quotient of 612 and 3.

4. -77 - -60 =

5. (-3) (-2) (2) =

6. $\dfrac{36 \div -9}{-8 \div -4} =$

Helpful Hints	Use what you have learned to solve problems. If necessary, refer to the examples on the previous page.

S1. $\dfrac{32 \div -4}{-8 \div 2} =$

S2. $\dfrac{5 \cdot -6}{-12 \div -4} =$

1. $\dfrac{2 \cdot -4}{-12 \div -6} =$

2. $\dfrac{-8 \text{ x } -5}{15 \div -3} =$

3. $\dfrac{4 \cdot -9}{-18 \div -3} =$

4. $\dfrac{-4 \cdot -20}{-32 \div 4} =$

5. $\dfrac{-54 \div -9}{-2 \cdot -3} =$

6. $\dfrac{40 \div -5}{-20 \div 5} =$

7. $\dfrac{10 \cdot -3}{-3 \cdot -5} =$

8. $\dfrac{32 \div -2}{-2 \cdot -2} =$

9. $\dfrac{-55 \div 11}{-25 \div -5} =$

10. $\dfrac{-42 \div 7}{-12 \div -4} =$

1.

2.

3.

4.

5.

6.

7.

8.

9.

10.

Score

Problem Solving	Elena scored a total of 475 points on five tests. What was her average score?

Review Exercises

1. 77 + 99 + 73 + 62 =

2. 7,165
      769
   + 963

3.   308
     x 206

4.   72,172
   - 15,564

5. 6 x 1,096 =

6. 6,102 - 1,765 =

Helpful Hints	Use what you have learned to solve the following problems. Remember to be careful with positive and negative signs.

S1. (-5) • 3 (-4) x 6 =

S2. $\dfrac{20 \div -2}{-25 \div 5} =$

1. 15 • -6 =

2. 6 x -5 • 4 =

3. (-3) (2) (-4) =

4. $\dfrac{-75}{-5}$

5. 32 ÷ -4 =

6. -224 ÷ -4 =

7. (-3) (-2) (6) =

8. $\dfrac{-30 \div -5}{2 \cdot -3} =$

9. $\dfrac{20 \text{ x } -3}{-50 \div -10} =$

10. $\dfrac{36 \div -9}{-16 \div 4} =$

1.

2.

3.

4.

5.

6.

7.

8.

9.

10.

Score

Problem Solving	Fariba had test scores of 80, 84, 96, and 100. What was her average score?

Review Exercises

1. $3\overline{)76}$

2. $\dfrac{1250}{5}$

3. $41\overline{)361}$

4. $41\overline{)765}$

5. $28\overline{)349}$

6. $28\overline{)2106}$

Helpful Hints	Use what you have learned to solve the following problems. Sometimes it is helpful to review the examples from previous pages if you need help.

S1. $-3 \cdot 4 \cdot -4 \cdot 3 =$

S2. $\dfrac{15 \cdot -2}{-2 \cdot -3}$

1. $25 \cdot (-6) =$

2. $5 \cdot (-6)(-3) =$

3. $(5)(-4)(2)(-2) =$

4. $\dfrac{-125}{-5}$

5. $54 \div -6 =$

6. $-111 \div -3 =$

7. $\dfrac{(-3)(-6)}{-9}$

8. $\dfrac{60 \div -6}{-2 \cdot -5} =$

9. $\dfrac{-72 \div 9}{-20 \div -5} =$

10. $\dfrac{20 \times -3}{-4 \times -5} =$

1.

2.

3.

4.

5.

6.

7.

8.

9.

10.

Score

Problem Solving

The attendance at the Eagle's game last year was 42,728. This year 53,962 attended. What was the increase in attendance this year?

Review of All Integer Operations

1. $-8 + 5 =$

2. $8 + -5 =$

3. $-8 + -5 =$

4. $-6 + -8 + 17 =$

5. $-35 + 19 + 23 + -33 =$

6. $6 - 8 =$

7. $4 - -8 =$

8. $-3 - 7 =$

9. $-12 - 16 =$

10. $18 - 19 =$

11. $4 \cdot -15 =$

12. $-3 \cdot -27 =$

13. $3 (-6) (-4) =$

14. $(-2) \cdot 5 (-6) \cdot 2 =$

15. $-27 \div 9 =$

16. $-234 \div -3 =$

17. $\dfrac{-256}{-8}$

18. $\dfrac{-24 \div 2}{18 \div 3}$

19. $\dfrac{8 \cdot (-4)}{-20 \div -5}$

20. $\dfrac{-30 \cdot -2}{-60 \div -10}$

1.	
2.	
3.	
4.	
5.	
6.	
7.	
8.	
9.	
10.	
11.	
12.	
13.	
14.	
15.	
16.	
17.	
18.	
19.	
20.	

Final Review of All Integer Operations

1. $12 + \text{-}6 =$

2. $\text{-}12 + 6 =$

3. $\text{-}12 + \text{-}6 =$

4. $\text{-}6 + \text{-}9 + 21 =$

5. $65 + \text{-}32 + 15 + \text{-}50 =$

6. $9 - 12 =$

7. $15 - \text{-}16 =$

8. $\text{-}20 - 31 =$

9. $\text{-}23 - 26 =$

10. $16 - 80 =$

11. $\text{-}16 \cdot 7 =$

12. $\text{-}15 \cdot \text{-}10 =$

13. $3\,(\text{-}12)\,(\text{-}4) =$

14. $(\text{-}2)\,3\,(\text{-}5)\,3 =$

15. $\dfrac{\text{-}72}{\text{-}8}$

16. $\text{-}414 \div \text{-}3 =$

17. $\dfrac{884}{\text{-}4}$

18. $\dfrac{80 \div \text{-}8}{\text{-}25 \div 5}$

19. $\dfrac{\text{-}12 \cdot \text{-}3}{\text{-}2 \cdot 2}$

20. $\dfrac{75 \div \text{-}25}{\text{-}3 \div \text{-}1}$

1.	
2.	
3.	
4.	
5.	
6.	
7.	
8.	
9.	
10.	
11.	
12.	
13.	
14.	
15.	
16.	
17.	
18.	
19.	
20.	

Whole Numbers - Final Test

1. $\begin{array}{r} 356 \\ + \ 397 \\ \hline \end{array}$

2. $\begin{array}{r} 623 \\ 462 \\ 17 \\ + \ 816 \\ \hline \end{array}$

3. $6,502 + 916 + 15,989 =$

4. $6,667 + 3,444 + 755 + 16 =$

5. $7,093 + 675 + 97 + 768 =$

6. $\begin{array}{r} 916 \\ - \ 428 \\ \hline \end{array}$

7. $\begin{array}{r} 5,392 \\ - \ 1,768 \\ \hline \end{array}$

8. $7,053 - 4,289 =$

9. $5,000 - 3,296 =$

10. $7,008 - 999 =$

11. $\begin{array}{r} 87 \\ \times \ \ 4 \\ \hline \end{array}$

12. $\begin{array}{r} 7,132 \\ \times \ \ \ \ 4 \\ \hline \end{array}$

13. $\begin{array}{r} 45 \\ \times \ 36 \\ \hline \end{array}$

14. $\begin{array}{r} 392 \\ \times \ \ 47 \\ \hline \end{array}$

15. $\begin{array}{r} 743 \\ \times \ 247 \\ \hline \end{array}$

16. $4\overline{)626}$

17. $4\overline{)1428}$

18. $40\overline{)568}$

19. $30\overline{)8672}$

20. $18\overline{)1343}$

1.	
2.	
3.	
4.	
5.	
6.	
7.	
8.	
9.	
10.	
11.	
12.	
13.	
14.	
15.	
16.	
17.	
18.	
19.	
20.	

Integers - Final Test

1. 7 + -5 =

2. -7 + 5 =

3. -7 + -5 =

4. -7 + 4 + -3 + 6 =

5. -52 + 17 + 22 + -21 =

6. 7 - 9 =

7. 4 - -7 =

8. -7 - 12 =

9. -12 - 16 =

10. 14 - 19 =

11. 4 • (-8) =

12. -12 • -19 =

13. 4 (-5) (-6) =

14. -2 • 3 • (-7) =

15. -63 ÷ 9 =

16. -560 ÷ -5 =

17. $\dfrac{-136}{-8}$

18. $\dfrac{32 \div -2}{-16 \div -4}$

19. $\dfrac{-8 \cdot -5}{-15 \div 3}$

20. $\dfrac{-40 \cdot -2}{-20 \div 2}$

1.	
2.	
3.	
4.	
5.	
6.	
7.	
8.	
9.	
10.	
11.	
12.	
13.	
14.	
15.	
16.	
17.	
18.	
19.	
20.	

Solutions

PAGE 4
Review Exercises:
1. 357
2. 627
3. 13
4. 739
5. 16
6. 49
S1. 819
S2. 933
1. 113
2. 783
3. 1,000
4. 1,072
5. 1,497
6. 687
7. 1,168
8. 120
9. 1,673
10. 412
Problem Solving: 134 students

PAGE 5
Review Exercises:
1. 449
2. 731
3. 83
4. 1,093
5. 1,178
6. 944
S1. 1,285
S2. 734
1. 1,264
2. 135
3. 1,767
4. 675
5. 1,913
6. 1,551
7. 329
8. 1,180
9. 288
10. 1,578
Problem Solving: $220

PAGE 6
Review Exercises:
1. 191
2. 1,225
3. 2,080
4. 1,906
5. 137
6. 62
S1. 7,619
S2. 44,027
1. 10,395
2. 24,708
3. 3,628,991
4. 7,497,988
5. 71,078
6. 94,180
7. 7,697
8. 105,598
9. 118,112
10. 899,935
Problem Solving: 36,283,995

PAGE 7
Review Exercises:
1. 885
2. 53,203
3. 134,680
4. 33,104,468
5. 211,628
6. 10,562,592
S1. 100,463
S2. 6,301,851
1. 78,734
2. 8,976,984
3. 12,915
4. 86,835
5. 20,422
6. 1,166,094
7. 212,299
8. 14,362,605
9. 97,625
10. 42,865
Problem Solving: 8,044 students

PAGE 8
Review Exercises:
1. 154
2. 43,939
3. 19,731,174
4. 1,615
5. 1,370,727
6. 337
S1. 384
S2. 2,181
1. 315
2. 571
3. 5,395
4. 2,775
5. 648
6. 5,098
7. 716
8. 3,389
9. 6,470
10. 491
Problem Solving: 97 students

PAGE 9
Review Exercises:
1. 1,883
2. 278
3. 6,057
4. 2,126
5. 4,861
6. 41,989,515
S1. 559
S2. 7,366
1. 473
2. 175
3. 1,885
4. 4,159
5. 4,667
6. 6,279
7. 11,874
8. 1,517
9. 2,369
10. 2,928
Problem Solving: $58

PAGE 10
Review Exercises:
1. 1,199
2. 6,766
3. 3,451
4. 8,692
5. 1,263
6. 2,656
S1. 245
S2. 433
1. 23
2. 529
3. 2,667
4. 462
5. 7,242
6. 3,668
7. 1,337
8. 2,216
9. 6,678
10. 7,485
Problem Solving: 291 seats

PAGE 11
Review Exercises:
1. 224
2. 4,214
3. 277
4. 4,249
5. 3,979
6. 4,232
S1. 4,546
S2. 6,033
1. 823
2. 2,428
3. 664
4. 4,803
5. 1,649
6. 42,458
7. 68,296
8. 4,494
9. 443
10. 47,192
Problem Solving: $16,500

PAGE 12
Review Exercises:
1. 830
2. 51,656
3. 1,969
4. 4,691
5. 4,236
6. 1,827
S1. 24,907
S2. 2,674
1. 632
2. 461
3. 13,049
4. 2,211
5. 4,303
6. 165
7. 9,979
8. 992
9. 778
10. 152,371
Problem Solving: $95

PAGE 13
Review Exercises:
1. 44
2. 73,212
3. 1,251
4. 64,738
5. 2,474
6. 79,922
S1. 5,655
S2. 52,811
1. 676
2. 5,764
3. 10,808
4. 27,891
5. 198
6. 344
7. 1,793
8. 67,384
9. 4,509
10. 37,046
Problem Solving: 8603 people

PAGE 14
Review Exercises:
1. 234
2. 1,052
3. 2,335
4. 148
5. 7,216
6. 40,117
S1. 1,578
S2. 19,524
1. 201
2. 444
3. 2,562
4. 25,284
5. 24,448
6. 43,536
7. 49,252
8. 15,396
9. 52,572
10. 34,848
Problem Solving: 2,555 days

PAGE 15
Review Exercises:
1. 1,372
2. 21,448
3. 2,803
4. 4,765
5. 874
6. 3,052
S1. 43,404
S2. 72,056
1. 4,368
2. 81,108
3. 166,082
4. 45,792
5. 87,570
6. 28,036
7. 21,144
8. 56,084
9. 142,368
10. 74,574
Problem Solving: 152 points

Solutions

PAGE 16
Review Exercises:
1. 4,350
2. 27,108
3. 15,846
4. 2,010
5. 3,327
6. 1,462
S1. 1,032
S2. 13,038
1. 4,346
2. 782
3. 3,995
4. 10,900
5. 16,800
6. 33,182
7. 592
8. 4,324
9. 34,335
10. 15,180
Problem Solving: 875 desks

PAGE 17
Review Exercises:
1. 2,839
2. 22,015
3. 1,728
4. 10,575
5. 15,726
6. 5,563
S1. 10,452
S2. 15,664
1. 1,300
2. 1,288
3. 2,530
4. 13,824
5. 18,848
6. 35,700
7. 9,476
8. 12,168
9. 4,550
10. 44,384
Problem Solving: 1200 girls

PAGE 18
Review Exercises:
1. 1,512
2. 11,396
3. 14,000
4. 6,255
5. 154
6. 4,164
S1. 64,719
S2. 380,824
1. 77,805
2. 33,702
3. 73,688
4. 183,092
5. 224,802
6. 107,600
7. 175,577
8. 67,808
9. 411,000
10. 162,504
Problem Solving: 158,600 cars

PAGE 19
Review Exercises:
1. 11,853
2. 824
3. 13,032
4. 1,058
5. 7,890
6. 370,304
S1. 99,900
S2. 317,100
1. 649,602
2. 371,500
3. 140,505
4. 489,240
5. 662,592
6. 229,272
7. 88,914
8. 427,000
9. 692,804
10. 612,381
Problem Solving: 784 students

PAGE 20
Review Exercises:
1. 2,448
2. 864
3. 25,334
4. 777
5. 4,452
6. 11,016
S1. 10,650
S2. 260,525
1. 185
2. 4,230
3. 16,422
4. 4,176
5. 15,435
6. 70,016
7. 10,890
8. 459,900
9. 20,130
10. 125,356
Problem Solving: 10,080 minutes

PAGE 21
Review Exercises:
1. 726
2. 3,135
3. 6,329
4. 1,498
5. 15,862
6. 182,000
S1. 30,035
S2. 19,085
1. 2,020
2. 42,012
3. 62,412
4. 7,680
5. 45,234
6. 307,824
7. 2,160
8. 70,704
9. 47,905
10. 158,704
Problem Solving: 101 people

PAGE 12

Review Exercises:
1. 307
2. 1,735
3. 88,625
4. 6,917
5. 2,992
6. 61,800
S1. 4 r1
S2. 9 r4
1. 24 r2
2. 5 r7
3. 13
4. 18 r3
5. 12
6. 15 r4
7. 9 r5
8. 24
9. 16 r1
10. 23 r1
Problem Solving: 18 pencils

PAGE 23

Review Exercises:
1. 8 r1
2. 17 r4
3. 6 r3
4. 106
5. 1155
6. 1551
S1. 18 r1
S2. 4 r3
1. 9 r2
2. 15
3. 11 r5
4. 48 r1
5. 8 r4
6. 5 r4
7. 15 r2
8. 24 r3
9. 8 r1
10. 13
Problem Solving: $78,000

PAGE 24

Review Exercises:
1. 6 r2
2. 13 r4
3. 4,486
4. 3,682
5. 3,772
6. 222
S1. 261 r1
S2. 45 r4
1. 324
2. 209 r1
3. 133 r2
4. 121 r1
5. 114 r4
6. 141 r4
7. 312
8. 121 r2
9. 258 r2
10. 89 r5
Problem Solving: 36 packages

PAGE 25

Review Exercises:
1. 19
2. 12 r5
3. 123 r1
4. 74 r4
5. 132
6. 34 r2
S1. 187 r1
S2. 94 r1
1. 114 r4
2. 78 r6
3. 97
4. 122
5. 127
6. 41 r2
7. 133 r1
8. 99
9. 155 r3
10. 101 r3
Problem Solving: 5,400 miles

PAGE 26

Review Exercises:
1. 23
2. 133 r2
3. 10,011
4. 5,373,870
5. 552,600
6. 773
S1. 3017 r1
S2. 670 r2
1. 4,617 r1
2. 3,681
3. 1,136 r5
4. 448
5. 1,533 r1
6. 1,780 r3
7. 12,247
8. 12,181
9. 10,309
10. 4,402 r4
Problem Solving: $3,310

PAGE 27

Review Exercises:
1. 175 r1
2. 201 r2
3. 376 r2
4. 1,235
5. 2,208 r4
6. 6,332
S1. 662 r4
S2. 2,795 r1
1. 1,374 r1
2. 345 r1
3. 1,112 r6
4. 887 r2
5. 1551 r3
6. 1,002
7. 5,574 r1
8. 3,186 r3
9. 15,263 r1
10. 4,714 r4
Problem Solving: 44 miles

PAGE 28
Review Exercises:
1. 782
2. 25,368
3. 11,102
4. 125
5. 193 r5
6. 119,507
S1. 300 r1
S2. 1,080 r3
1. 80 r2
2. 1,700
3. 2,103 r1
4. 445 r6
5. 254 r2
6. 1,200
7. 2,002 r1
8. 973 r8
9. 904 r1
10. 804 r1
Problem Solving: 12,500 sheets

PAGE 29
Review Exercises:
1. 2,459
2. 319,774
3. 85 r5
4. 1,000 r3
5. 1,525 r1
6. 281 r2
S1. 123
S2. 1,462
1. 61 r2
2. 17 r4
3. 202 r1
4. 1,001 r5
5. 1,622 r4
6. 349 r2
7. 1,677 r7
8. 2,877 r2
9. 11,872 r2
10. 3,071
Problem Solving: 556 containers

PAGE 30
Review Exercises:
1. 119 r4
2. 1,142
3. 226,800
4. 1,001 r2
5. 233 r1
6. 16,253
S1. 5 r26
S2. 143 r12
1. 3 r28
2. 7 r26
3. 8 r38
4. 11 r59
5. 56 r2
6. 69 r16
7. 224 r12
8. 101 r6
9. 94 r26
10. 59 r36
Problem Solving: 256 boxes

PAGE 31
Review Exercises:
1. 34 r1
2. 1,002 r5
3. 1,577
4. 10,944
5. 5,132
6. 2,041
S1. 2 r37
S2. 126 r52
1. 19 r16
2. 9 r7
3. 77 r2
4. 246 r16
5. 81 r54
6. 142 r48
7. 199 r26
8. 44 r6
9. 27 r49
10. 241 r26
Problem Solving: 768 students

PAGE 32
Review Exercises:
1. 1,046
2. 5,768
3. 250
4. 12 r9
5. 5 r17
6. 131 r36
S1. 21 r1
S2. 8 r23
1. 22 r34
2. 22 r1
3. 12 r1
4. 9 r15
5. 24 r10
6. 31 r15
7. 12 r15
8. 11 r26
9. 44 r12
10. 20 r26
Problem Solving: 14 gallons

PAGE 33
Review Exercises:
1. 11,410
2. 162,000
3. 206
4. 2,244
5. 16 r26
6. 81 r82
S1. 22 r10
S2. 7 r10
1. 13 r30
2. 22 r18
3. 13 r2
4. 36 r4
5. 32 r9
6. 24 r18
7. 10 r46
8. 31 r2
9. 34 r24
10. 17 r36
Problem Solving: 185 seats

PAGE 34
Review Exercises:
1. 3 r8
2. 13 r13
3. 8 r3
4. 13 r10
5. 8 r19
6. 14 r13
S1. 3 r71
S2. 19 r4
1. 5 r7
2. 5 r82
3. 54 r15
4. 8 r2
5. 5 r3
6. 63 r4
7. 19 r3
8. 19 r31
9. 20 r1
10. 43 r4
Problem Solving: 900 parts

PAGE 35
Review Exercises:
1. 78
2. 17,332
3. 1,472
4. 5,175
5. 250,800
6. 388,367
S1. 29 r21
S2. 37 r16
1. 31 r14
2. 32 r5
3. 30 r2
4. 20 r16
5. 29 r28
6. 49
7. 43 r13
8. 19 r31
9. 63 r4
10. 8 r10
Problem Solving: $53.00

PAGE 36
Review Exercises:
1. 365
2. 467
3. 30,658
4. 1,765
5. 2,969
6. 1,347
S1. 215 r5
S2. 207 r11
1. 125 r20
2. 148 r15
3. 214 r17
4. 165 r11
5. 298 r10
6. 306 r1
7. 32 r62
8. 80 r5
9. 142
10. 750 r6
Problem Solving: 440 miles

PAGE 37
Review Exercises:
1. 35 r1
2. 872 r1
3. 107 r42
4. 42 r27
5. 12 r2
6. 8 r33
S1. 48 r5
S2. 212 r11
1. 94 r22
2. 77 r35
3. 114 r14
4. 94 r28
5. 116 r5
6. 51 r54
7. 317 r17
8. 163 r19
9. 401 r1
10. 43 r3
Problem Solving:
11 cartons, 27 left

PAGE 38
Review Exercises:
1. 107,083
2. 4,329
3. 155,935
4. 218,255
5. 5,110
6. 7,392
S1. 95 r21
S2. 55 r 21
1. 38 r1
2. 331 r3
3. 279 r2
4. 16 r7
5. 7 r76
6. 41 r72
7. 25 r16
8. 12 r2
9. 54 r50
10. 253
Problem Solving: 51 students

PAGE 39
Review Exercises:
1. 34
2. 654 r2
3. 9 r9
4. 45 r17
5. 11 r25
6. 73 r11
S1. 96 r16
S2. 77 r17
1. 19 r2
2. 325 r5
3. 778
4. 16 r7
5. 6 r36
6. 84 r56
7. 5 r16
8. 12 r50
9. 23 r48
10. 130 r28
Problem Solving: $960

PAGE 40
1. 1,109
2. 2,899
3. 8,1387
4. 10,675
5. 8,474
6. 345
7. 2,815
8. 6,285
9. 4,012
10. 6,285
11. 261
12. 45,794
13. 3,087
14. 61,484
15. 113,702
16. 141 r2
17. 282 r5
18. 21 r27
19. 150 r17
20. 44 r26

PAGE 41
1. 1,846
2. 89,844
3. 32,904
4. 8,638
5. 26,297
6. 435
7. 8,215
8. 44,499
9. 2,643
10. 55,807
11. 2,156
12. 63,090
13. 2,523
14. 38,142
15. 771,363
16. 58 r6
17. 232
18. 5 r39
19. 142 r28
20. 594 r10

PAGE 42
Review Exercises:
1. 1,840
2. 6,894
3. 3,783
4. 233 r6
5. 5 r17
6. 17 r39
S1. 6
S2. -23
1. 9
2. -18
3. -16
4. -27
5. 7
6. 38
7. -180
8. -29
9. -70
10. -20
Problem Solving: 6 buses

PAGE 43
Review Exercises:
1. -14
2. 5
3. -78
4. 148,131
5. 12 r25
6. 6 r5
S1. 9
S2. -185
1. -9
2. -936
3. -37
4. -141
5. 37
6. -215
7. -901
8. -99
9. -641
10. 154
Problem Solving: $31

PAGE 44
Review Exercises:
1. 21,063
2. 1,257
3. 58,398
4. -15
5. -188
6. 37
S1. -2
S2. -5
1. 3
2. -3
3. -6
4. 13
5. -6
6. -8
7. -33
8. -9
9. -26
10. -89
Problem Solving: 89

PAGE 45
Review Exercises:
1. 24 r7
2. 9 r24
3. 21 r25
4. 62 r3
5. 2
6. -72
S1. -6
S2. 9
1. -7
2. -22
3. -55
4. 14
5. -34
6. 14
7. -24
8. -108
9. 33
10. -57
Problem Solving: 24 packages

Solutions

PAGE 46
Review Exercises:
1. -28
2. 8
3. -15
4. 241
5. 1,229
6. 17,292
S1. -13
S2. 1
1. 15
2. -3
3. 8
4. -31
5. 66
6. 34
7. -20
8. -18
9. -93
10. 31
Problem Solving: -8°

PAGE 47
Review Exercises:
1. -9
2. -10
3. -30
4. -42
5. 20
6. -16
S1. -28
S2. 112
1. 38
2. -8
3. 12
4. -46
5. 71
6. -79
7. -14
8. 48
9. -434
10. -16
Problem Solving: $195

PAGE 48
Review Exercises:
1. -16
2. -2
3. 40
4. 21
5. -4
6. -107
S1. -46
S2. 22
1. -105
2. 15
3. -21
4. 60
5. -23
6. 9
7. -29
8. 19
9. 76
10. -924
Problem Solving: -34

PAGE 49
Review Exercises:
1. -185
2. -171
3. 361
4. 156
5. -1,014
6. -22
S1. -965
S2. 157
1. -65
2. 72
3. 71
4. -53
5. 41
6. -95
7. -119
8. -25
9. -68
10. -738
Problem Solving: 21°

PAGE 50
Review Exercises:
1. 766
2. 749
3. 8,208
4. 27 r23
5. -64
6. 10
S1. 48
S2. -136
1. 174
2. -80
3. 504
4. -594
5. -1440
6. -171
7. 56
8. 384
9. -272
10. 646
Problem Solving: $330

PAGE 51
Review Exercises:
1. 42
2. -156
3. -300
4. -12
5. 29
6. 15
S1. -180
S2. 320
1. 110
2. -110
3. 432
4. -420
5. -1,600
6. 375
7. -1,320
8. 720
9. -576
10. 2,400
Problem Solving: 105 feet

Solutions

PAGE 52

Review Exercises:
1. -7
2. 2
3. 42
4. 252
5. -23
6. 36
S1. 64
S2. -30
1. -60
2. -126
3. 360
4. 180
5. -12
6. -120
7. -48
8. 192
9. 252
10. 440

Problem Solving: 3,680 miles

PAGE 53

Review Exercises:
1. 24
2. -105
3. 44
4. -84
5. -36
6. -24
S1. 378
S2. -180
1. 120
2. -105
3. -144
4. 450
5. -200
6. 120
7. -336
8. 360
9. 252
10. 216

Problem Solving: 375 cows

PAGE 54

Review Exercises:
1. -1
2. 2
3. 5
4. -75
5. 18
6. 60
S1. -3
S2. 15
1. -18
2. -48
3. 21
4. -51
5. -26
6. -48
7. -8
8. -17
9. 13
10. 19

Problem Solving: 12 packs

PAGE 55

Review Exercises:
1. 3 r2
2. 199 r3
3. 14 r56
4. 6 r34
5. 5 r26
6. 15 r9
S1. -32
S2. 6
1. -18
2. -81
3. -23
4. 707
5. -72
6. 22
7. 16
8. -5
9. 26
10. 667

Problem Solving:
   59 miles per hour

PAGE 56

Review Exercises:
1. -96
2. 12
3. 24
4. -50
5. -4
6. 6
S1. -1
S2. -12
1. 2
2. 3
3. -4
4. -18
5. -3
6. 2
7. 1
8. -3
9. -3
10. 2

Problem Solving: $1.75

PAGE 57

Review Exercises:
1. 21
2. 25,368
3. 204
4. -17
5. 12
6. -2
S1. 2
S2. -10
1. -4
2. -8
3. -6
4. -10
5. 1
6. 2
7. -2
8. -4
9. -1
10. -2

Problem Solving: 95

Solutions

PAGE 58
Review Exercises:
1. 311
2. 8,897
3. 63,448
4. 56,608
5. 6,576
6. 4,337
S1. 360
S2. 2
1. -90
2. -120
3. 24
4. 15
5. -8
6. 56
7. 36
8. -1
9. -12
10. 1
Problem Solving: 90

PAGE 59
Review Exercises:
1. 25 r1
2. 250
3. 8 r33
4. 18 r27
5. 12 r13
6. 75 r6
S1. 144
S2. -5
1. -150
2. 90
3. 80
4. 25
5. -9
6. 37
7. -2
8. -1
9. -2
10. -3
Problem Solving: 11,234

PAGE 60
1. -3
2. 3
3. -13
4. 3
5. -26
6. -2
7. 12
8. -10
9. -28
10. -1
11. -60
12. 81
13. 72
14. 120
15. -3
16. 78
17. 32
18. -2
19. -8
20. 10

PAGE 61
1. 6
2. -6
3. -18
4. 6
5. -2
6. -3
7. 31
8. -51
9. -49
10. -64
11. -112
12. 150
13. 144
14. 90
15. 9
16. 138
17. -221
18. 2
19. -9
20. -1

PAGE 62
1. 753
2. 1,918
3. 23,407
4. 10,882
5. 8,633
6. 488
7. 3,624
8. 2,764
9. 1,704
10. 6,009
11. 348
12. 28,528
13. 1,620
14. 18,424
15. 183,521
16. 156 r2
17. 357
18. 14 r8
19. 289 r2
20. 74 r11

PAGE 63
1. 2
2. -2
3. -12
4. 0
5. -34
6. -2
7. 11
8. -19
9. -28
10. -5
11. -32
12. 228
13. 120
14. 42
15. -7
16. 112
17. 17
18. -4
19. -8
20. -8

Math Notes